D1412909

'He can explain the complexities of cosmological physics with an engaging combination of clarity and wit . . . His is a brain of extraordinary power'
Observer

'To follow such a fine mind as it exposes such great problems is an exciting experience'
Sunday Times

'One of the most brilliant scientific minds since Einstein'
Daily Express

'He has certainly done more to make the idea of a big bang and the beginning of the universe coherent and thinkable than any physicist since Einstein'
Newsday

'A high-priest of physics, one of a handful of theorists who may be on the verge of reading God's mind'
Los Angeles Times

'Genius, unique, tragic and triumphant . . . Hawking takes us through the evolution of modern thinking on cosmology, from Aristotle and Copernicus, through Galileo and Newton, to Einstein and, indeed, Hawking himself'
Sydney Morning Herald

Books by Stephen Hawking:

A Brief History of Time
Black Holes and Baby Universes and Other Essays
The Illustrated A Brief History of Time
The Universe in a Nutshell
My Brief History

With Leonard Mlodinow:
A Briefer History of Time
The Grand Design

With Lucy Hawking:
George's Secret Key to the Universe
George's Cosmic Treasure Hunt
George and the Big Bang
George and the Unbreakable Code
George and the Blue Moon

Black Holes:
The BBC Reith Lectures

Stephen Hawking

With an introduction and notes by
BBC News Science Editor David Shukman

BANTAM BOOKS

LONDON · TORONTO · SYDNEY · AUCKLAND · JOHANNESBURG

TRANSWORLD PUBLISHERS
61–63 Uxbridge Road, London W5 5SA
www.transworldbooks.co.uk

Transworld is part of the Penguin Random House group of companies
whose addresses can be found at global.penguinrandomhouse.com

'Do Black Holes Have No Hair?' first broadcast by BBC Radio 4
on 26 January 2016.
'Black Holes Ain't As Black As They Are Painted' first broadcast by
BBC Radio 4 on 2 February 2016.

First published by arrangement with the BBC in Great Britain
in 2016 by Bantam Books
an imprint of Transworld Publishers

Copyright © Stephen Hawking 2016

Stephen Hawking has asserted his right under the Copyright,
Designs and Patents Act 1988 to be identified as the author of this work.

The animations and illustrations were produced by Cognitive
(wearecognitive.com) for BBC Radio 4.

The BBC Radio 4 logo is a trade mark of the British Broadcasting
Corporation and is used under licence.
BBC Radio 4 © 2011.

A CIP catalogue record for this book
is available from the British Library.

ISBN
9780857503572

Typeset in 12/16 pt Adobe Caslon by
Jouve (UK), Milton Keynes
Printed and bound in Great Britain by Clays Ltd, Bungay, Suffolk.

Penguin Random House is committed to a sustainable
future for our business, our readers and our planet. This book is
made from Forest Stewardship Council® certified paper.

MIX
Paper from
responsible sources
FSC® C018179

3 5 7 9 10 8 6 4

CONTENTS

INTRODUCTION
by David Shukman

Everything about Stephen Hawking is a source of fascination: the plight of a genius trapped in an ailing body; the hint of a smile brightening a face in which only a single muscle can move; the distinctively robotic voice inviting us to share the exhilaration of discovery as his mind roams through the strangest corners of the Universe.

Against all the odds, this remarkable figure has transcended the usual boundaries of science. His book *A Brief History of Time* sold a staggering ten million copies. Cameo roles in popular comedy shows, invitations to the

White House and a well-received movie about his life have confirmed him as a celebrity. He has achieved nothing less than becoming the most famous scientist in the world.

In the 1960s, he was given two years to live when he was diagnosed with motor neurone disease. But more than half a century later he is still researching, writing, travelling and regularly appearing in the news. His daughter Lucy, explaining this extraordinary drive, describes him as 'enormously stubborn'.

Whether through the pain of his personal story or his ability to enthuse, Hawking captures the imagination. He recently warned that humankind faced a series of disasters of its own making – from global warming to artificially engineered viruses – and an article reporting his words was the most-read on the BBC website that day.

It is a terrible irony that such a great communicator cannot have a normal conversation. For interviews, the questions have to be sent in advance. Some years ago, his staff warned me not to attempt small talk because his answers even to the briefest questions take so long to compose. In the excitement of meeting him, however, I could not resist blurting out: 'How are you?' – and then had to wait guiltily for his reply. He was fine.

In his Cambridge office, a board is covered with equations. Mathematics of the most rarefied kind is the currency of cosmology. But Stephen Hawking's unique contribution to scientific research is to harness the approaches of apparently very different specialisms: most famously, he was the first to investigate the vast realm of space using scientific techniques devised to study the tiny particles inside atoms.

His colleagues in this fiendishly complex field might fear that their work can never be made intelligible to the public. Yet striving to reach a wider audience is a Hawking trademark. In this year's BBC Reith Lectures, he rose to the challenge of summarizing a lifetime's insights into black holes in two fifteen-minute talks. And for those who are curious but perplexed, or enthralled by the ideas but nervous about the science, I have added notes at key points to offer a helping hand.

DO BLACK HOLES HAVE NO HAIR?

Broadcast 26 January 2016

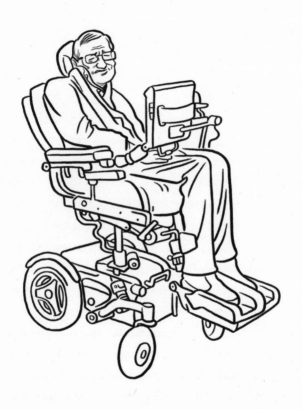

It is said that fact is sometimes stranger than fiction, and nowhere is that more true than in the case of black holes. Black holes are stranger than anything dreamed up by science-fiction writers, but they are firmly matters of science fact. The scientific community was slow to realize that massive stars could collapse in on themselves, under their own gravity, and to consider how the objects left behind would behave. Albert Einstein even wrote a paper in 1939 claiming that stars could not collapse under gravity because matter could not be compressed beyond a certain point. Many scientists shared Einstein's gut feeling. The principal exception was the

American scientist John Wheeler, who in many ways is the hero of the black hole story. In his work in the 1950s and 1960s, he emphasized that many stars would eventually collapse, and pointed out the problems that possibility posed for theoretical physics. He also foresaw many of the properties of the objects which collapsed stars become – that is, black holes.

DS: The phrase 'black hole' is simple enough, but it's hard to imagine one out there in space. Think of a giant drain with water spiralling down into it. Once anything slips over the edge – what is called the 'event horizon' – there is no way back. Because black holes are so powerful, even light gets sucked in, so we can't actually see them. But scientists know they exist because they rip apart stars that get too close to them and because they can send tremors through space. It was a collision between two black holes more than a billion years ago that triggered what are called 'gravitational waves', the recent detection of which was a hugely significant scientific achievement.

During most of the life of a normal star, over many billions of years, it will support itself against its own gravity by thermal pressure, caused by nuclear processes which convert hydrogen into helium.

DS: NASA describes stars as rather like pressure-cookers. The explosive force of nuclear fusion inside them creates outward pressure which is constrained by gravity pulling everything inwards.

Eventually, however, the star will exhaust its nuclear fuel. The star will now contract. In

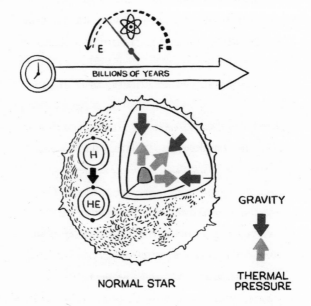

BILLIONS OF YEARS

H

HE

NORMAL STAR

GRAVITY

THERMAL PRESSURE

some cases, it may be able to support itself as a 'white dwarf' star. However, Subrahmanyan Chandrasekhar showed in 1930 that the maximum mass of a white dwarf star is about 1.4 times that of the sun. A similar maximum mass was calculated by Soviet physicist Lev Landau for a star made entirely of neutrons.

DS: White dwarfs and neutron stars were once suns and have since burned up their fuel. With no force working to inflate them, nothing can stop their gravitational pull from shrinking them, and they have become some of the densest objects in the universe. But in the league-table of stars these ones are relatively small, which means they lack the gravitational strength to collapse completely. So, most interesting to Stephen Hawking and others is what happens to the very largest stars when they reach their end of their lives.

What, then, would be the fate of those countless stars with greater mass than a white dwarf or neutron star when they had exhausted their nuclear fuel? The problem was investigated by Robert Oppenheimer, of later atom bomb fame. In a couple of papers in 1939, with George Volkoff and Hartland Snyder, he showed that such a star could not be supported by outward pressure; and that, if you take pressure out of the calculation, a uniform spherically systematic symmetric star would contract to a single point of infinite density. Such a point is called a singularity.

DS: A singularity is what you end up with when a giant star is compressed to an unimaginably small point. This concept has been a defining theme in Stephen Hawking's career. It refers not only to the end of a star but also to a far more fundamental idea about the starting-point for the formation of

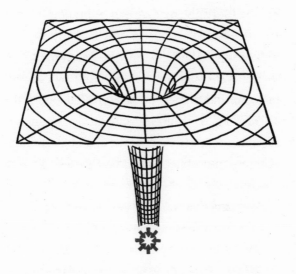

the entire universe. It was Hawking's mathematical work on this that earned him global recognition.

All our theories of space are formulated on the assumption that space-time is smooth and nearly flat, so they break down at the singularity, where the curvature of space-time is infinite. In fact, the singularity marks the end of time itself. That is what Einstein found so objectionable.

DS: Einstein's Theory of General Relativity says that objects distort the space-time around them. Picture a bowling-ball lying on a trampoline, changing the shape of the material and causing smaller objects to slide towards it. This is how the effect of gravity is explained. But if the curves in space-time become deeper and deeper, and eventually infinite, the usual rules of space and time cease to apply.

Then the Second World War intervened. Most scientists, including Robert Oppenheimer, switched their attention to nuclear physics, and the issue of gravitational collapse was largely forgotten. Interest in the subject revived with the discovery of distant objects called quasars.

DS: Quasars are the brightest objects in the universe, and possibly the most distant detected so far. The name is short for 'quasi-stellar radio sources' and they are believed to be discs of matter swirling around black holes.

The first quasar, 3C273, was discovered in 1963. Many other quasars were soon discovered. They were bright, despite being very distant. Nuclear processes could not account for their energy output, because they release only a tiny amount of their rest mass as pure energy. The only alternative was gravitational energy, released by gravitational collapse. Thus gravitational collapses of stars were rediscovered.

It was already clear that a uniform spherical star would contract to a point of infinite density, a singularity. The Einstein equations don't work at a singularity. This means that at this point of infinite density, one can't predict the future, which in turn implies that something strange could happen whenever a star collapsed. We wouldn't be affected by the breakdown of prediction if the singularities were naked, that is, if they were not shielded from the outside.

DS: A 'naked' singularity is a theoretical scenario in which a star collapses but an event horizon does not form around it – so the singularity would be visible.

When John Wheeler introduced the term 'black hole' in 1967, it replaced the earlier name 'frozen star'. Wheeler's coinage emphasized that the remnants of collapsed stars are of interest in their own right, independently of how they were formed. The new name caught on quickly. It suggested something dark and mysterious. But the French, being French, saw a more risqué meaning. For years, they resisted the name *trou noir*, claiming it was obscene. But that was a bit like trying to stand against *le weekend*, and other Franglais. In the end, they had to give in. Who can resist a name that is such a winner?

From the outside, you can't tell what is inside a black hole. You can throw television sets, diamond rings, or even your worst enemies into a black hole, and all the black hole will remember is the total mass, the state

of rotation and the electric charge. John Wheeler is known for expressing this principle as 'a black hole has no hair'. To the French, this just confirmed their suspicions.

A black hole has a boundary, called the event horizon. This is where gravity is just strong enough to drag light back and prevent it escaping. Because nothing can travel faster than light, everything else will get dragged back also. Falling through the event horizon is a bit like going over Niagara Falls in a canoe. If you are above the falls, you can get away if you paddle fast enough, but once you are over the edge, you are lost. There's no way back. As you get nearer the falls, the current gets faster. This means it pulls harder on the front of the canoe than the back. There's a danger that the canoe will be pulled apart. It is the same with black holes. If you fall towards a black hole feet first, gravity will pull harder on your feet than your head, because they are nearer the black hole. The result is you will be stretched out longways, and squashed in sideways. If the black hole has a mass of a few times our sun's you will be torn apart and made into spaghetti before you reach the horizon. However, if you fall towards a much larger black hole, with a mass of a million times the sun's, you'll reach the horizon without difficulty. So, if you want to explore the inside of a black hole, make sure you choose a big one. There is a black hole with a mass of about four million times that of the sun at the centre of our Milky Way galaxy.

DS: Scientists believe that there are huge black holes at the centre of virtually all galaxies – a remarkable thought, given how recently these features were confirmed in the first place.

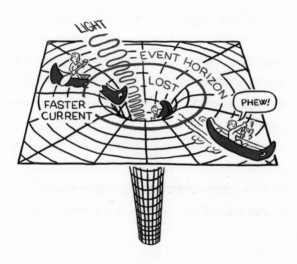

Although you wouldn't notice anything particular as you fell into a black hole, someone watching you from a distance would never see you cross the event horizon. Instead, you would appear to slow down, and hover just outside. Your image would get dimmer and dimmer, and redder and redder, until you were effectively lost from sight. As far as the outside world is concerned, you would be lost for ever.

DS: With no light escaping from the black hole, there is no way that anyone watching from a distance could actually witness your descent. In space, no one can hear you scream; and in a black hole, no one can see you disappear.

A dramatic advance in our understanding of these mysterious phenomena came with a mathematical discovery in 1970. This was that the surface area of the event horizon, the boundary zone around a black hole, has the property that it always increases when additional matter or radiation falls into the black hole. This property suggests that there is a resemblance between the area of the event horizon of a black hole, and conventional Newtonian physics, specifically the concept of entropy in thermodynamics. Entropy can be regarded as a measure of the disorder of a system, or equivalently, as a lack of knowledge of its precise state. The famous Second Law of Thermodynamics says that entropy always increases with time. The 1970 discovery was the first hint of this crucial connection.

DS: Entropy means the tendency for anything that has order to become more disordered as time passes – so, for example, bricks neatly stacked to form a wall (low entropy) will eventually end up in an untidy heap of dust (high entropy). And this process is described by the Second Law of Thermodynamics.

Although the existence of a connection between entropy and the area of the event horizon was clear, it was not obvious to us how the area could be identified as the entropy of a black hole itself. What would be meant by the entropy of a black hole? The crucial suggestion was made in 1972 by Jacob Bekenstein, a graduate student at Princeton University who later worked at the Hebrew University of Jerusalem. It goes like this. When a black hole is created by gravitational collapse, it rapidly settles down to a stationary state, which is characterized by only three parameters: the mass, the angular momentum (state of rotation) and the electric charge. Apart from these three properties, the black hole preserves no other details of the object that has collapsed.

This theorem has implications for information, in the cosmologist's sense of information: the idea that every particle and every force in the universe has an implicit answer to a yes–no question.

DS: Information, in this context, means all the details of every particle and every force associated with an object. The more disordered something is – the higher its entropy – the more information is needed to describe it. As the physicist and broadcaster Jim Al-Khalili puts it, a well-shuffled pack of cards has higher entropy than an unshuffled one and therefore its description requires far more explanation, or information.

EVENT HORIZON

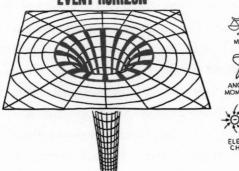

MASS

ANGULAR
MOMENTUM

ELECTRIC
CHARGE

Bekenstein's theorem implies that a large amount of information is lost in a gravitational collapse. For example, the final black-hole state is independent of whether the body that collapsed was composed of matter or antimatter, or whether it was spherical or highly irregular in shape. In other words, a black hole of a given mass, angular momentum and electric charge could have been formed by the collapse of any one of a large number of different configurations of matter – including any one of a large number of different types of star. Indeed, if quantum effects are left aside, the number of potential configurations would be infinite, since the black hole could have been formed by the collapse of a cloud of an indefinitely large number of particles, of indefinitely low mass. But could the number of configurations really be infinite? This is where quantum effects come in.

The uncertainty principle of quantum mechanics implies that only particles with a wavelength smaller than that of the black hole itself could form a black hole. That means the range of potential wavelengths would be limited: it could not be infinite.

DS: The uncertainty principle, conceived by the famous German physicist Werner Heisenberg in the 1920s, states that we can never locate or predict the precise position of the smallest particles. So, at what is called the quantum scale, there is a fuzziness in nature, very unlike the precisely ordered universe described by Isaac Newton.

It therefore appears that the number of configurations that could form a black hole of

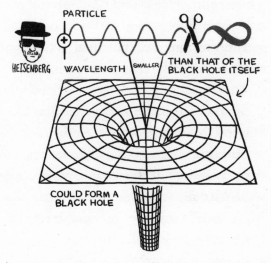

UNCERTAINTY PRINCIPLE

PARTICLE

HEISENBERG

WAVELENGTH SMALLER THAN THAT OF THE BLACK HOLE ITSELF

COULD FORM A BLACK HOLE

a given mass, angular momentum and electric charge, although very large, may also be finite. Jacob Bekenstein suggested that from this finite number one could derive the entropy of a black hole. This would be a measure of the amount of information that was irretrievably lost during the collapse when a black hole was created.

The apparently fatal flaw in Bekenstein's suggestion was that if a black hole has a finite entropy that is proportional to the area of its event horizon, it also ought to have a finite temperature, which would be proportional to its surface gravity. This would imply that a black hole could be in equilibrium with respect to thermal radiation, at some temperature other than zero. Yet according to classical concepts, no such equilibrium is possible, since the black hole would absorb any thermal radiation that fell on it, but by definition would not be able to emit anything in return. It cannot emit anything. It cannot emit heat.

> *DS: If information is lost, which is apparently what is happening in a black hole, there should be some release of energy – but that flies in the face of the theory that nothing comes out of black holes.*

This is a paradox. And it's one to which I am going to return in my next lecture, when I'll be exploring how black holes challenge the most basic principle about the predictability of the universe, and the certainty of history – and asking what would happen if you ever got sucked into one.

DS: So Stephen Hawking has taken us on a scientific journey from Einstein's claim that stars could not collapse, through the acceptance of the reality of black holes, to a collision of theories over how these weird features exist and function.

BLACK HOLES AIN'T AS BLACK AS THEY ARE PAINTED

Broadcast 2 February 2016

In my previous lecture I left you on a cliffhanger: a paradox about the nature of black holes, the incredibly dense objects created by the collapse of stars. One theory suggested that black holes with identical qualities could be formed from an infinite number of different types of stars. Another suggested that the number of possible types could be finite. This is a problem of information, that is, the idea that every particle and every force in the universe contains an implicit answer to a yes–no question.

Because 'black holes have no hair', as the scientist John Wheeler put it, one can't tell from the outside what is inside a black hole, apart from its mass, state of rotation and electric charge. This means that a black hole contains a lot of information that is hidden from the outside world. If the amount of information hidden inside a black hole depends on the size of the hole, one would expect from general principles that the black hole would have a temperature, and would glow like a piece of hot metal. But that was impossible, because, as everyone knew, nothing could get out of a black hole. Or so it was thought.

This paradox persisted until early in 1974, when I was investigating what the behaviour of matter in the vicinity of a black hole would be, according to quantum mechanics.

DS: Quantum mechanics is the science of the extremely small and it seeks to explain the behaviour of the tiniest particles. These do not act according to the laws that govern the movements of much bigger objects like planets, laws that were first framed by Isaac Newton. Using the science of the very small to study the very large was one of Stephen Hawking's pioneering achievements.

To my great surprise I found that the black hole seemed to emit particles at a steady rate. Like everyone else at that time, I accepted the dictum that a black hole could not emit anything. I therefore put quite a lot of effort into trying to get rid of this embarrassing effect. But the more I thought about it, the more it refused to go away, so that in the end I had to accept it. What finally convinced me it was a real physical process was that the wavelengths of the outgoing particles were precisely thermal. My calculations predicted that a black hole creates and emits particles and radiation just as if it were an ordinary hot body, with a temperature that is proportional to its surface gravity and inversely proportional to its mass.

> *DS: These calculations were the first to show that a black hole need not be a one-way street to a dead end. Not surprisingly, the emissions suggested by the theory became known as Hawking Radiation.*

Since that time, the mathematical evidence that black holes emit thermal radiation has been confirmed by a number of other people taking various different approaches. One way to understand these emissions is as follows. Quantum mechanics implies that the whole of space is filled with pairs of virtual particles and antiparticles, which are constantly materializing in pairs, separating, and then coming together again and annihilating each other.

> DS: *This concept hinges on the idea that a vacuum is never totally empty. According to the uncertainty principle of quantum mechanics, there is always the chance that particles may come into existence, however briefly. And this would always involve pairs of particles, with opposite characteristics, appearing and disappearing.*

These particles are called 'virtual' because, unlike real particles, they cannot be observed directly with a particle detector. Their indirect effects can nonetheless be measured, and their existence has been confirmed by a small shift, called the Lamb shift, which they produce in the level of light-spectrum

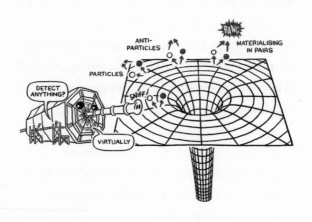

energy emitted by excited hydrogen atoms. Now, in the presence of a black hole, one member of a pair of virtual particles may fall into the hole, leaving the other member without the partner necessary for mutual annihilation. The forsaken particle or anti-particle may fall into the black hole after its partner, but it may also escape to infinity, where it appears to be radiation emitted by the black hole.

DS: The key here is that the formation and disappearance of these particles normally passes unnoticed. But if the process happens right on the edge of a black hole, one of the pair may be dragged in while the other is not. The particle that escapes would then look as if it were being 'spat out' by the black hole.

A black hole of the mass of the sun would leak particles at such a slow rate that the process would be impossible to detect. However, there could be much smaller 'mini' black holes with the mass of say, a mountain. A mountain-sized black hole would give off X-rays and gamma rays at a rate of about ten million megawatts, enough to power the entire world's electricity supply. It wouldn't be easy, however, to harness a mini black hole. You couldn't keep it in a power station, because it would drop through the floor and end up at the centre of the earth. If we had such a black hole, about the only way to keep hold of it would be to have it in orbit around the earth.

People have searched for mini black holes of this mass, but so far have not found any. This is a pity, because if they had I would have got a

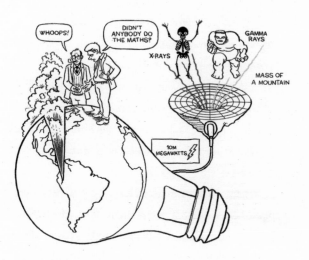

Nobel Prize! Another possibility, however, is that we might be able to create micro black holes in the extra dimensions of space-time.

DS: These 'extra dimensions' refer to something beyond the three dimensions that we are all familiar with in our everyday lives, plus the fourth dimension of time. The idea arose as part of an effort to explain why gravity is so much weaker than other forces such as magnetism – maybe it's also having to operate in parallel dimensions.

According to some theories, the universe we experience is just a four-dimensional surface in a ten- or eleven-dimensional space. The movie *Interstellar* gives some idea of what this would be like. We wouldn't see these extra dimensions because light wouldn't propagate through them, but only through the four

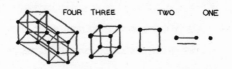

FOUR THREE TWO ONE

TEN OR ELEVEN DIMENSIONS

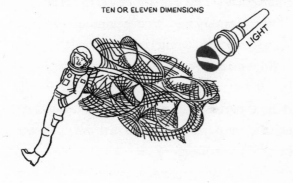

LIGHT

dimensions of our universe. Gravity, however, would affect the extra dimensions and would be much stronger there than in our universe. This would make it much easier for a little black hole to form in the extra dimensions. It might be possible to observe this at the LHC, the Large Hadron Collider, at CERN in Switzerland. This consists of a circular tunnel, 27 kilometres long. Two beams of particles travel round this tunnel in opposite directions, and are made to collide. Some of the collisions might create micro black holes. These would radiate particles in a pattern that would be easy to recognize. So I might get a Nobel Prize after all!

DS: The Nobel Prize in Physics is awarded when a theory is 'tested by time' which in practice means confirmed by hard evidence. For example, Peter

Higgs was one of the scientists who, back in the 1960s, suggested the existence of a particle that would give other particles their mass. Nearly fifty years later, two different detectors at the Large Hadron Collider spotted signs of what had become known as the Higgs Boson. It was a triumph of science and engineering, of clever theory and hard-won evidence; and as a result Peter Higgs and François Englert, a Belgian scientist, were jointly awarded the prize. No physical proof has yet been found of Hawking Radiation and some scientists suggest that it will be too difficult to detect. Still, with black holes being studied in ever closer detail, some day confirmation may come.

As particles escape from a black hole, the hole will lose mass, and shrink. This will increase the rate of emission of particles. Eventually, the black hole will lose all its mass and disappear. What then happens to all the particles and unlucky astronauts that fell into the black hole? They can't just re-emerge when the black hole disappears. It seems that the information about what fell in is lost, apart from the total amount of mass, the amount of rotation and the electric charge. But if information *is* lost, this raises a serious problem that strikes at the heart of our understanding of science.

For more than two hundred years we have believed in scientific determinism, that is, that the laws of science determine the evolution of the universe. This principle was formulated by Pierre-Simon Laplace, who said that if we know the state of the universe at one time, the laws of science will determine it at all future and past times. Napoleon is said to have asked Laplace how God fitted into this picture, and Laplace to have replied: 'Sire, I have not needed that hypothesis.' I don't think that Laplace was claiming that God didn't exist – just that he doesn't intervene to break the laws of science. That must be the position of every scientist. A scientific law is not a scientific law if it only holds when some supernatural being decides to let things run and not intervene.

In Laplace's determinism, one needed to know the positions and speeds of all particles at one time, in order to predict the future. But we need also to take into account the uncertainty principle, articulated by Walter Heisenberg in 1923, which lies at the heart of quantum mechanics.

This holds that the more accurately you know the positions of particles, the less accurately you can know their speeds, and vice versa. In other words, you can't know both the positions and the speeds accurately. How, then, can you predict the future accurately? The answer is that although one can't predict the positions and speeds separately, one can predict what is called the 'quantum state'. This is something from which both positions and speeds can be calculated to a certain degree of accuracy. We would still expect the universe to be deterministic, in the sense that if we knew the quantum state of the universe at one time, the laws of science should enable us to predict it at any other time.

DS: What began as an explanation of what happens at an event horizon has deepened into an exploration of some of the most important philosophical themes in science – from the clockwork

world of Newton to the laws of Laplace to the
uncertainties of Heisenberg – and of the points at
which they are challenged by the mystery of black
holes. Essentially, while according to Einstein's
Theory of General Relativity information entering
a black hole is destroyed, quantum theory says it
cannot be broken down.

If information were lost in black holes, we wouldn't be able to predict the future, because a black hole could emit any collection of particles. It could emit a working television set, or a leather-bound volume of the complete works of Shakespeare, though the chance of such exotic emissions is very small. It might seem that it wouldn't matter very much if we couldn't predict what comes out of black holes. There aren't any black holes near us. But it is a matter of principle.

If determinism, the predictability of the universe, breaks down with black holes, it could break down in other situations. Even worse, if determinism breaks down, we can't be sure of our past history either. The history books and our memories could just be illusions. It is the past that tells us who we are; without it, we lose our identity.

It was therefore very important to determine whether information really was lost in black holes or whether, in principle, it could be recovered. Many scientists felt that information should not be lost, but no one could suggest a mechanism by which it could be preserved. The arguments went on for years. Finally, I found what I think is the answer. It depends on the idea of Richard Feynman that instead of one single history there are many different possible histories, each with its own probability. In this case, there are two kinds of history. In one, there is a black hole, into which particles can fall; in the other, there is no black hole.

The point is that from the outside, one can't be certain whether there is a black hole or not. So there is always a chance that there isn't a black hole. This possibility is enough to preserve the information, but the information is not returned in a very useful form. It is like burning an encyclopaedia. Information is not lost if you keep all the smoke and ashes, but it is difficult to read. The scientist Kip Thorne and I had a bet with another physicist, John Preskill, that information would be lost in black holes. When I discovered how information could be preserved, I conceded the bet. I gave John Preskill an encyclopaedia. Maybe I should have just given him the ashes.

> *DS: In theory, and with a purely deterministic view of the universe, you could burn an encyclopaedia and then reconstitute it – if you knew the characteristics and position of every atom making up every molecule of ink and paper and kept track of them all at all times.*

Currently I'm working with my Cambridge colleague Malcolm Perry and Andrew Strominger from Harvard on a new theory based on a mathematical idea called supertranslations, with the aim of explaining the mechanism by which information is returned out of the black hole. According to our theory, the information is encoded on the horizon of the black hole. Watch this space!

DS: Since the Reith Lectures were recorded, Professor Hawking and his colleagues have published a paper which makes a mathematical case that information can be stored in the event horizon. The theory hinges on information being transformed into a two-dimensional hologram in a process known as supertranslation. The paper, entitled 'Soft Hair on Black Holes', offers a highly revealing glimpse into the esoteric language of this field – as the abstract reproduced at the end of this lecture shows – and the challenge that scientists face in trying to explain it.

What does this tell us about whether it is possible to fall into a black hole and come out in another universe? The existence of alternative histories with and without black holes suggests this might be possible. The hole would need to be large, and if it were rotating, it might have a passage to another universe. But you couldn't come back to our universe. So although I'm keen on space flight, I'm not going to try that.

> DS: If a black hole is rotating, then its heart may not consist of a singularity in the sense of an infinitely dense point. Instead, there may be a singularity in the form of a ring. And that leads to speculation about the possibility of not only falling into a black hole but also travelling through one. This would mean leaving the universe as we know it. And Stephen Hawking concludes with a tantalizing thought: that there may be something on the other side.

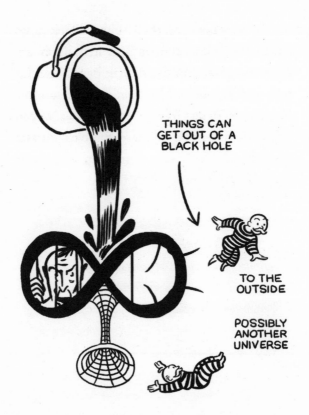

THINGS CAN
GET OUT OF A
BLACK HOLE

TO THE
OUTSIDE

POSSIBLY
ANOTHER
UNIVERSE

My message here, then, is that black holes aren't as black as they are painted. They are not the eternal prisons once envisaged. Things can get out of a black hole, both in this universe and possibly to another. So, if you feel you are in a black hole, don't give up: there is a way out!

Soft Hair on Black Holes

Stephen W. Hawking[†], Malcolm J. Perry[†] and Andrew Strominger[*]

[†]*DAMTP, Centre for Mathematical Sciences,*
University of Cambridge, Cambridge, CB3 0WA UK
[*] *Center for the Fundamental Laws of Nature,*
Harvard University, Cambridge, MA 02138, USA

Abstract

It has recently been shown that BMS supertranslation symmetries imply an infinite number of conservation laws for all gravitational theories in asymptotically Minkowskian spacetimes. These laws require black holes to carry a large amount of soft (*i.e.* zero-energy) supertranslation hair. The presence of a Maxwell field similarly implies soft electric hair. This paper gives an explicit description of soft hair in terms of soft gravitons or photons on the black hole horizon, and shows that complete information about their quantum state is stored on a holographic plate at the future boundary of the horizon. Charge conservation is used to give an infinite number of exact relations between the evaporation products of black holes which have different soft hair but are otherwise identical. It is further argued that soft hair which is spatially localized to much less than a Planck length cannot be excited in a physically realizable process, giving an effective number of soft degrees of freedom proportional to the horizon area in Planck units.

arXiv:1601.00921v1 [hep-th] 5 Jan 2016

STEPHEN HAWKING is regarded as one of the most brilliant theoretical physicists since Einstein.

In 1963, aged twenty-one and a graduate student at Cambridge University, Stephen Hawking contracted motor neurone disease and was given two years to live. Yet he went on to become a brilliant researcher and Professorial Fellow at Gonville and Caius College and then held the post of Lucasian Professor of Mathematics and Theoretical Physics, the chair held by Isaac Newton in 1663, for thirty years. Professor Hawking is now Director of Research for the Centre for Theoretical Cosmology at the University of Cambridge. He has over a dozen honorary degrees, and was awarded the Companion of Honour in 1989. He is a fellow of the Royal Society and a Member of the US National Academy of Science.

Professor Hawking is the author of *A Brief History of Time*, which was an international bestseller. His other bestselling books for the general reader include *A Briefer History of Time*, the essay collection *Black Holes and Baby Universes and Other Essays*, *The Universe in a Nutshell* and *The Grand Design*. He lives in Cambridge.

DAVID SHUKMAN is BBC News Science Editor and has reported on scientific and environmental issues since 2003. His assignments have ranged from the launch of the last American space shuttle to the discoveries of the Large Hadron Collider. David, a regular contributor to the BBC's News at Ten, is the author of three books.

If you'd like to read more by Stephen Hawking ...

A Brief History of Time: From the Big Bang to Black Holes

Professor Hawking's internationally acclaimed masterpiece begins by reviewing the great theories of the cosmos from Newton to Einstein, before delving into the secrets at the heart of space and time – from the Big Bang to black holes, via spiral galaxies and string theory. First published in 1988, *A Brief History of Time* remains a staple of the scientific canon, and its succinct and clear language continues to introduce millions to the wonders of the universe.

Black Holes and Baby Universes and Other Essays

This first collection of shorter writings, on subjects ranging from the warmly personal to the coolly scientific, reveals Stephen Hawking as scientist, as man, as concerned world citizen and – always – as rigorous and imaginative thinker. Whether he is remembering his first experience of nursery school, puncturing the arrogance of those who think science can best be understood only by other scientists and should be left to them, exploring the origins and the future of the universe, or reflecting on the phenomenon of *A Brief History of Time*, Stephen Hawking writes with the wit, clarity and directness that make him one of our age's greatest communicators.

The Universe in a Nutshell

This lavishly illustrated book brings us to the cutting edge of theoretical physics, where truth is often stranger than fiction. Here Stephen Hawking turns to the major break-throughs that occurred in the decade following the release of *A Brief History of Time*, guiding us on his own quest to uncover the secrets of the universe – from supergravity to supersymmetry, from quantum theory to M-theory, from holography to duality. In this most exciting intellectual adventure he seeks 'to combine Einstein's General Theory of Relativity and Richard Feynman's idea of multiple histories into one complete unified theory that will describe everything that happens in the universe'.

The Grand Design: New Answers to the Ultimate Questions of Life (with Leonard Mlodinow)

When and how did the universe begin? Why are we here? Is the apparent 'grand design' of our universe evidence for a benevolent creator who set things in motion? Or does science offer another explanation? In this most recent major work, written in collaboration with the American physicist and writer Leonard Mlodinow, we are presented with the latest scientific thinking about the mysteries of the universe in language marked by both brilliance and simplicity. Model-dependent realism, the multiverse, the top-down theory of cosmology, the unified M-theory – all are revealed in this succinct and startling illustrated guide to the discoveries that are altering our understanding and threatening some of our most cherished belief systems.